J
595.1
STO

Stone, Lynn M.

Leeches.

006894

$14.60

DATE			
302			
MAY 13	4		
112			
NOV 04			
207			
NOV 12			
302			
MAR 12	7		
302			
MAY 16			

LEECHES

CREEPY CRAWLERS

Lynn Stone

The Rourke Book Co., Inc.
Vero Beach, Florida 32964

PHOTO CREDITS
Title page, page 7 © James H. Robinson; cover and page 4 ©
Michael Cardwell; pages 8 and 12-13 © Breck P. Kent; page 10 ©
Dick Todd/Photo USA; pages 15 and 17 © Gordon Wiltsie; Page
18 © James P. Rowan; page 21 © James M. Gale

Library of Congress Cataloging-in-Publication Data

Stone, Lynn M.
    Leeches / by Lynn M. Stone.
        p. cm. — (Creepy crawlers)
    Includes index.
    Summary: Describes what leeches look like, where they live,
what they eat, how they travel, who their relatives are, and how
some of them draw blood.
    ISBN 1-55916-162-0
    1. Leeches—Juvenile literature. [1. Leeches.]
I. Title II. Series: Stone, Lynn M.   Creepy crawlers
QL391.A6S74   1995
595.1'45—dc20   95–16560
                                                      CIP
                                                      AC

Printed in the USA

TABLE OF CONTENTS

LEECHES

As a swimmer in a pond, you may find companions you would not choose—bloodsuckers.

Actually, bloodsuckers are leeches—soft, boneless animals that are related to earthworms. Unlike earthworms, however, leeches can be nasty creatures with a taste for blood.

A leech's companionship won't do any permanent damage, at least in the United States or Canada. A biting leech is just doing what it's made to do.

A blood-sucking leech makes itself a meal from a swimmer's flesh

WHAT LEECHES LOOK LIKE

Imagine an earthworm flattened into a pear shape. That almost describes a leech. Trouble is, a leech's shape is like a drop of water—it changes as it moves. Some leeches may be pear-shaped, then suddenly flex themselves into the shape of a bone or keyhole.

Leeches are equipped with a small sucker on the underside of the head. They have a larger sucker under the rear end of the body.

The largest **species** (SPEE sheez), or kind, of leech in the United States is about a foot long.

An aquatic leech shows one of its suckers

RELATIVES

Scientists call boneless animals, like **slugs** (SLUHGZ), worms, and leeches, **invertebrates** (in VERT uh brayts). Leeches belong to a group of invertebrates called **annelids** (AN nel lidz).

Slugs are not annelids, and many worms are not. Earthworms are annelids, so they are fairly close cousins of the leeches. Those family ties, however, don't prevent leeches from gobbling up earthworms.

American leeches have a special fondness for eating their earthworm cousins

WHAT LEECHES EAT

American leeches will travel nearly a mile from ponds, where they normally live, to eat earthworms on land. Leeches also eat other small invertebrates, including their own young!

Certain kinds of leeches feed on flesh and body fluids of animals much larger than themselves. Other leeches dine on snails and slugs, which they swallow whole.

This leech may eat its own young when they hatch

An aquatic leech, seen from its top side, swims over a pond bottom

BLOODSUCKERS

Not all leeches are bloodsuckers. Certain species of leeches, like vampire bats, do drink blood. Freshwater American leeches often draw blood from the hind legs of turtles.

A blood-sucking leech prepares for its meal by first attaching its rear sucker to an animal. A leech uses its mouth, inside the front sucker, to chew a wound.

A hiker removes a leech from his leg

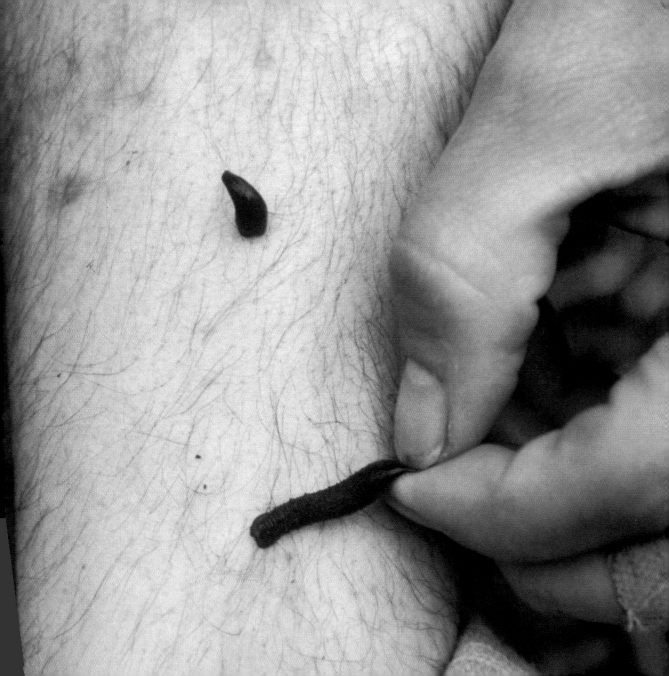

DRAWING BLOOD

The leech sucks blood through the wound. As it draws blood, the leech changes shape from flat to fat. The once-streamlined leech fills like a water balloon.

Amazingly, a leech can take in blood that weighs 10 times more than the empty leech did. After its big drink, the leech drops off its victim and rests.

A blood-sucking leech is a master of gentle biting. A crocodile it's not! A leech's victim often has no idea it is sharing its blood.

With its suckers firmly in place, a leech fills on human blood

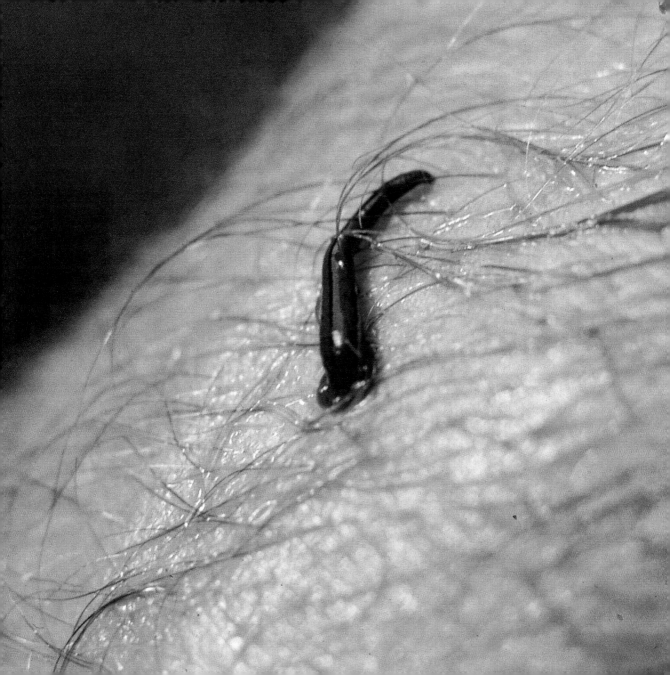

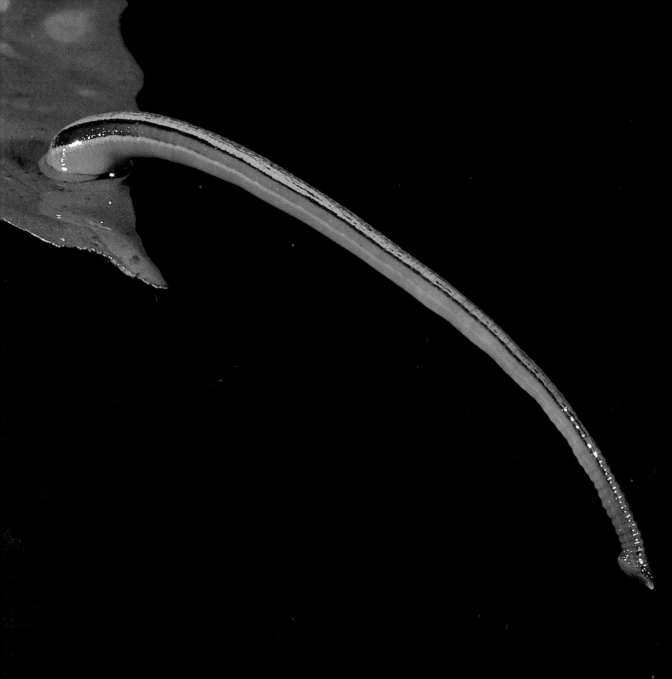

HOW LEECHES TRAVEL

Most of the leech's cousins, such as earthworms, travel by crawling. Leeches, though, aren't really built for crawling. Land leeches, for example, travel with stepping movements.

While one sucker grasps a surface, the rest of the leech's body loops forward—up, over, and past the first sucker. The leech attaches the second sucker and releases the first. Then the leech lifts the first sucker and steps past the second.

Aquatic (uh KWAHT ihk), or water, leeches travel mostly by swimming from place to place.

With one sucker attached, a leech swings 19
outward, feeling for another perch

WHERE LEECHES LIVE

Leeches live almost anywhere they find fairly still water or moisture. Aquatic leeches live in oceans, ponds, lakes, marshes, water-filled ditches, and slow-moving streams.

Land leeches may live in damp soil or on plants. They are plentiful in the warm, wet tropical forests of the world, especially in Southeast Asia.

Most American leeches are aquatic, but they sometimes hunt on land. Aquatic leeches are very common in the upper half of the United States.

A colorful leech wriggles through a rain forest in Borneo

LEECHES AND PEOPLE

No one enjoys finding a leech on his or her arm. Leeches, though, aren't a major health problem in North America. Leeches are a problem in warm countries, like Sri Lanka.

Doctors once thought—wrongly—that leeches could cure illnesses by sucking out "bad" blood. Millions of leeches were used for that purpose in Europe.

Today doctors still use leeches—for a substance in their bodies. That substance, in medicine form, helps keep blood flowing properly in some human patients.

Glossary

annelids (AN nel lidz) — boneless animals whose bodies are made of ringed sections, or segments; earthworms, leeches, and their relatives

aquatic (uh KWAHT ihk) — of or related to water, such as an aquatic bird

invertebrates (in VERT uh brayts) — the simple, boneless animals, such as worms, snails, starfish, and slugs

slugs (SLUHGZ) — boneless animals that are basically snails without shells

species (SPEE sheez) — within a group of closely related animals, one certain kind, such as a *Siberian* tiger

INDEX